KU-034-323

This book belongs to:

....

OUR WONDERFUL WORLD

How this collection works

This collection includes seven fascinating non-fiction texts designed to tap into your child's interest in the wonderful world around them, from exotic plants and fabulous beaches, to coral reefs and African plains. These texts are packed full of fascinating information, with the same high-quality artwork and photos you would expect from any non-fiction book – but they are specially written so that your child can read them for themselves. They are carefully levelled and in line with your child's phonics learning at school.

It's very important for your child to have access to non-fiction as well as stories while they are learning to read. This helps them develop a wider range of reading skills, and prepares them for learning through reading. Most children love finding out about the world as they read – and some children prefer non-fiction to story books, so it's doubly important to make sure that they have opportunities to read both.

How to use this book

Reading should be a shared and enjoyable experience for both you and your child. Pick a time when your child is not distracted by other things, and when they are happy to concentrate for about 10 minutes. Choose one or two of the non-fiction texts for each session, so that they don't get too tired. Read the tips on the next page, as they offer ideas and suggestions for getting the most out of this collection.

Tips for reading non-fiction

STEP 1

Before your child begins reading one of the non-fiction texts, look together at the contents page for that particular text. What does your child think the text will be about? Do they know anything about this subject already? Briefly talk about your child's ideas, and remind them of anything they know about the topic if necessary. Look at the topic words and other notes for each text, and use the 'before reading' suggestions to help introduce the text to your child.

STEP 2

Point out some of the non-fiction features in the text – for example, the contents page, and any photographs. Talk about how the contents page helps you find the different parts of the text, and the photographs help show that this is a book about the real world rather than a story.

STEP 3

Ask your child to read the text aloud. Encourage them to stop and look at the pictures, and talk about what they are reading either during the reading session, or afterwards. Your child will be able to read most of the words in the text, but if they struggle with a word, remind them to say the sounds in the word from left to right and then blend the sounds together to read the whole word, e.g. *e-x-o-t-i-c, exotic*. If they have real difficulty, tell them the word and move on.

STEP 4

When your child has finished reading, talk about what they have found out. Which bits of the text did they like most, and why? Encourage your child to do some of the fun activities that follow each text.

CONTENTS

OXFORD
UNIVERSITY PRESS

Big Animal Vet

This text follows a vet who works on a wildlife reserve in Africa.
He travels around in a helicopter and helps animals who are unwell.

Before reading

Does your child know what a vet does? If necessary, explain that
they are a bit like an animal doctor.

Topic words

These words may be challenging to read but they are important
for the topic. Read them together and talk about what they mean.

helicopter – an aircraft with a big propeller at the top

hunt – to chase and kill animals for food

buffalo – a wild animal similar to a cow, with long horns

elephant – a very large animal with grey skin, big ears and
a long trunk

Tricky words

These words are common but your child might find them difficult
to read:

my, her, they, are, all

BIG ANIMAL VET

CONTENTS

Jillian Powell

Zebra

I am a vet and I help animals!
I set off in my helicopter.

A big cat bit this zebra.
I check her neck.

Big cats hunt zebras.

big cat

Buffalo

Then I go to check on a sick buffalo.

buffalo

This jab will help him.

Big Cats

Then I visit a big cat and her cubs.

cub

I check they are all well.

Elephant

I get back in the helicopter.
Then I spot a sick elephant!

I can help her get well.
I am glad I am a vet!

My Trip

Talk about it!

Would you like to be a vet? What would you enjoy about it?

Maze

Help the vet get to the sick buffalo.

Exotic Plant Shop

In this text, a boy visits a shop full of amazing exotic plants and learns about each of them.

Before reading

Talk about any plants your child knows from your house, garden or nearby area. What are they like?

Topic words

These words may be challenging to read but they are important for the topic. Read them together and talk about what they mean.

plant – a living thing that grows in soil

exotic – unusual and interesting

lithops – (say lith-ops) a small plant that looks like a stone

eat – to use something as food

Tricky words

These words are common but your child might find them difficult to read:

my, we, you

EXOTIC PLANT SHOP

CONTENTS

Becca Heddle

We can get an
exotic plant in
this shop.

21

This plant is a cactus.

It pricks me!

22

This plant has a cup. Insects drop in. The plant eats them.

This plant is a trap.

The trap shuts to eat insects.

Yuck!

My plant is the best!

Talk about it!

Would you like an exotic plant? What kind of plant would you choose?

Letter scramble

Unscramble the letters to make the plant names.

s t a c c u

p i l th o s

Things with Wings

This text looks at things which have wings and can fly, such as insects, bats and birds. It also looks at birds that cannot fly.

Before reading

What kinds of creatures have wings? See how many different kinds of winged creatures you and your child can think of.

Topic words

These words may be challenging to read but they are important for the topic. Read them together and talk about what they mean.

butterfly – an insect with a thin body and two pairs of large wings

fly – move through the air

birds – animals that have wings and feathers

talons – long, strong claws

Tricky words

These words are common but your child might find them difficult to read:

are, all

THINGS WITH WINGS

CONTENTS

Paul Shipton

Insects

This is a butterfly wing.

wings

A butterfly is an insect.

Lots of insects can fly.

Bats

This is a bat wing.

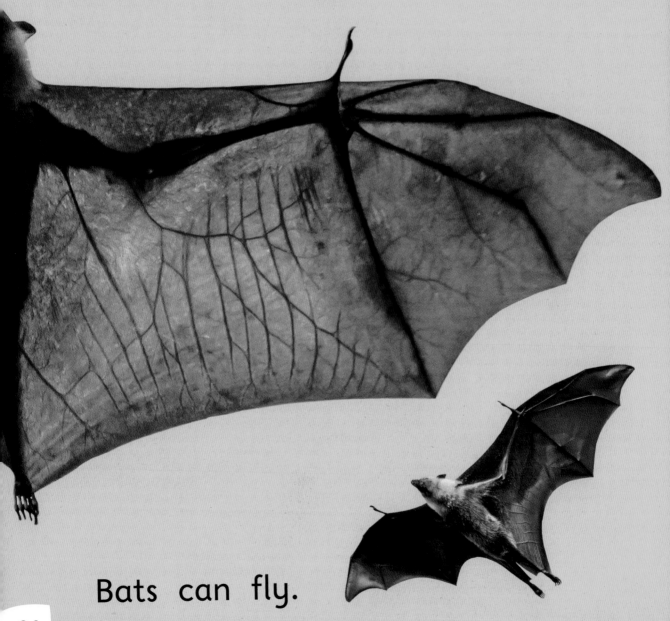

Bats can fly.

This bat can fly
to hunt moths.

Birds

This is a parrot wing.

Lots of birds
can fly.

This bird has big wings. It picks up fish with its talons.

talons

This bird has wings that are not big at all!

Birds That Cannot Fly

Not all birds can fly!

This bird swims with its wings.

This bird runs on its long legs.

Things with Wings

Insects

Bats

Birds

Birds that cannot fly

Talk about it!

What was your favourite thing with wings? Why?

Who can fly?

Which two of these creatures have wings but cannot fly?

41

Off to the Beach

This text explores four different beaches from around the world and shows some of the animals that live there.

Before reading

Has your child ever been to the beach? Talk about what beaches are like and what kinds of animals, birds and plants might live there.

Topic words

These words may be challenging to read but they are important for the topic. Read them together and talk about what they mean.

beach – an area of land next to water

mangrove – a swamp full of mangrove trees

coral – a sea creature which looks like a knobbly rock

ice – frozen water

Tricky words

This word is common but your child might find it difficult to read:

we

OFF TO THE BEACH

CONTENTS

Off we go!

Rob Alcraft

Beach

We set off in a ship.

We spot a beach.
It has cliffs and rocks.

Mangrove

This is a mangrove.
It has a lot of wet mud.

bat

mantis

moth

We spot a moth
and a bat!

Off we go!

Coral

This beach has sand.
The sun is hot.

coral

Ice

This beach has lots of ice!

We spot animals on the ice.
We spot fish as well.

cod

pup

Back we go!

Back We Go!

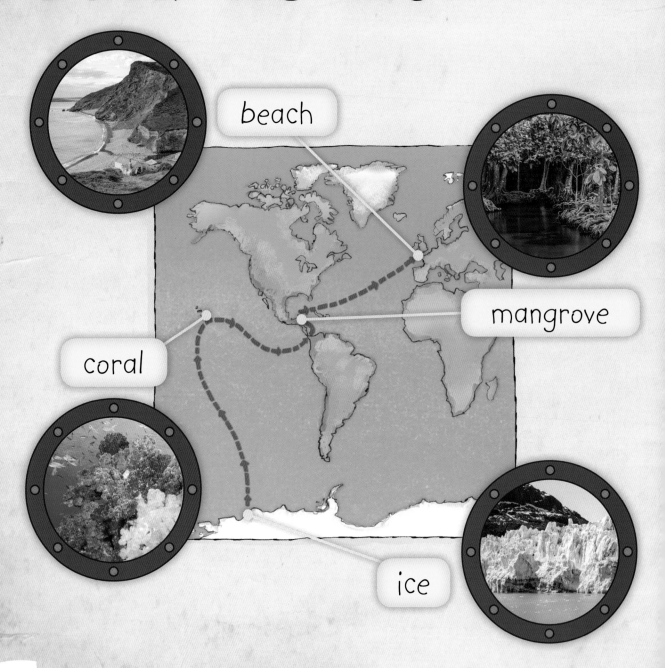

beach

mangrove

coral

ice

Talk about it!

Which of the beaches in this text would you like to visit? Why?

Spot the creatures

All of these creatures are in the text. Can you spot them? What are they called?

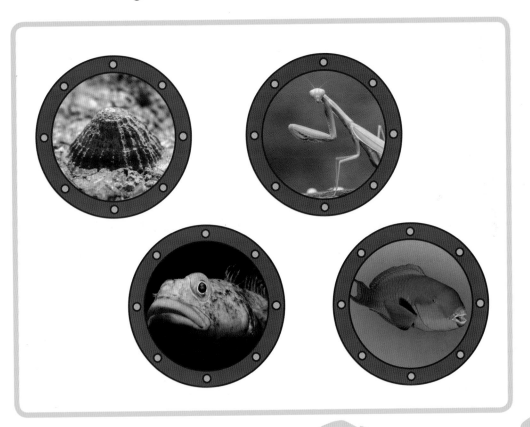

How Can I Help You?

This text looks at some living things which have a relationship that benefits them both.

Before reading

Explain to your child that this book is about animals who live together and help each other. How do they think one animal could help another animal?

Topic words

These words may be challenging to read but they are important for the topic. Read them together and talk about what they mean.

clean – not dirty; to remove dirt

pollen – a powder found inside flowers that helps plants to make seeds

coral – a sea creature which looks like a knobbly rock

starfish – a sea creature shaped like a star

Tricky words

These words are common but your child might find them difficult to read:

you, my, they, me, are

HOW CAN I HELP YOU?

CONTENTS

Becca Heddle

Zebra and Oxpecker

oxpecker

zebra

I need to get rid of ticks on my skin. They can harm me!

tick

I peck the ticks. They are my food.

oxpecker

It is good for us.

57

Shark and Remora

shark

I need to get my skin clean.

remora

I suck its skin to get my food. This cleans the skin!

remora

It is good for us.

Flower and Bee

I need food.

I need pollen for my seeds.

bee

I get food from the flower and I bring pollen.

pollen

flower

It is good for us.

Coral and Crab

crab

coral

I need to keep clean and stop starfish feeding on me.

I shelter in the coral and clean it. I fight off starfish.

starfish

crab

It is good for us.

Animals That Help

Talk about it!

Can you remember how the oxpecker helps me?

Match them up

Match the labels to the pictures.

tick

starfish

remora

pollen

How We See

This text looks at how different animals see and compares the eyesight of people, goats, horses and cats.

Before reading

Does your child know that different types of animals do not see the world in the same way?

Topic words

These words may be challenging to read but they are important for the topic. Read them together and talk about what they mean.

eye – the part of the body that we see with

horse – a large four-legged animal that people can ride on

newborn – very recently born

baby – a very young child or animal

Tricky words

These words are common but your child might find them difficult to read:

we, my, you

HOW WE SEE

CONTENTS

Kate Scott

Goat

This is my eye.

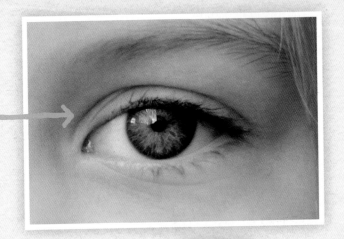

This is the eye of a goat.

I see this much.

A goat sees this much.

It is hard to creep up on a goat!

Horse

My eyes!

The eyes of a horse.

I see this.

A horse sees this!

A horse has a part it cannot see!
You *can* creep up on a horse.

Cat

My eye!

The eye of a cat.

I cannot see much at night
but a cat can!

At night,
I see this.

A cat
sees this!

Newborn Baby

My eye!

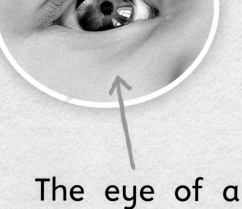

The eye of a newborn baby.

A newborn baby cannot
see things well.

I see this.

A newborn
baby sees this!

How We See

Me

Goat

Horse

Cat

Newborn baby

Talk about it!

Would you like to be able to see like a horse? Why, or why not?

Who am I?

Read the speech bubbles and name the animals.

I can *see* well at night. Who am I?

It is hard to creep up on me. Who am I?

I cannot *see* things well. Who am I?

You can creep up on me. Who am I?

Dive! Dive!

This text follows some people in a submarine as it dives deeper and deeper under the sea.

Before reading

What does your child think they might see if they dived down deep under the sea?

Topic words

These words may be challenging to read but they are important for the topic. Read them together and talk about what they mean.

submarine – a type of ship that can travel under the sea

sea – the large areas of salty water found all around the world

dive – go down into the water

metres – a way of measuring the length or depth of something

Tricky words

These words are common but your child might find them difficult to read:

we, me, are

DIVE! DIVE!

CONTENTS

Liz Miles

The Submarine

In a submarine we can look deep under the sea.

fish

This tells me how much air is in the submarine.

Going Down

The submarine is now 100 metres deep.

This deep, we can still see in the sunlight.

orca

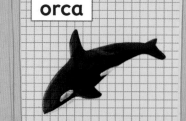

82

83

Deeper

Now the submarine is much deeper. It is darker and cooler.

I put the lights on.

shark

In the Dark

This deep, it is hard to see.

Look! That odd fish is an eel.

eel

On This Dive...

fish

orca

shark

eel

Talk about it!

Which of the sea creatures was the most interesting? Why?

Zoom out

Unscramble the letters and match the animal names to the pictures.

k a s h r

e l e

c a o r

i s h f

ACTIVITIES

Find the names!

Trace the lines to find the animal names.

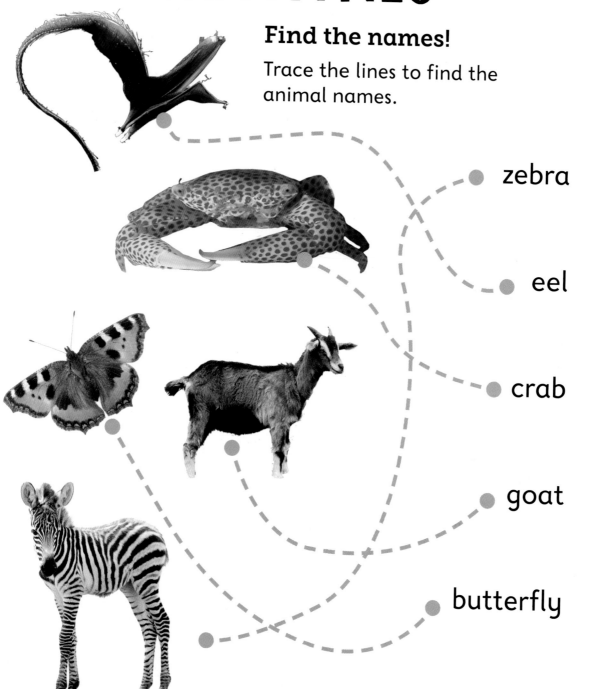

zebra

eel

crab

goat

butterfly

Hidden words!

Find the words in the grid.

i	a	s	h	a	r	k	t	u	s
n	w	u	r	n	e	e	w	b	s
s	a	n	a	c	c	o	r	a	l
e	f	f	g	i	r	n	b	c	a
c	d	l	e	r	t	t	o	l	c
t	h	o	r	s	e	h	r	h	a
f	x	w	b	d	f	g	c	k	c
s	o	e	l	e	p	h	a	n	t
s	o	r	e	m	o	r	a	l	u
r	a	k	u	s	r	o	y	o	s

cactus

coral

orca

horse

insect

elephant

shark

sunflower

remora

92

Spot the difference!

Look for the differences between the pictures.
Can you spot two differences for each picture?

Quick quiz

Can you answer the quiz question about each book?

Big Animal Vet

1 How does the vet travel around?

How We See

2 Which animal sees well in the dark?

Exotic Plant Shop

3 Why does the boy not want a lithops?

Off to the Beach

4 Name one kind of animal you could find near a mangrove.

Things with Wings

5 Name one animal that can fly, but isn't a bird.

Dive! Dive!

6 Which underwater animal do the children see first?

How Can I Help You?

7 Which creature helps the flower?

Answers: 1. in a helicopter; 2. cat; 3. he thinks they are dull; 4. one of bat, moth or mantis; 5. bat or butterfly; 6. fish; 7. bee

OXFORD
UNIVERSITY PRESS

Great Clarendon Street, Oxford, OX2 6DP, United Kingdom

Oxford University Press is a department of the University of Oxford. It furthers the University's objective of excellence in research, scholarship, and education by publishing worldwide. Oxford is a registered trade mark of Oxford University Press in the UK and in certain other countries

Big Animal Vet text © Oxford University Press 2016
Illustrations © Joe Todd-Stanton 2016

Exotic Plant Shop text © Becca Heddle 2016
Illustrations © Beatriz Castro 2016

Things with Wings text © Paul Shipton 2016

Off to the Beach text © Rob Alcraft 2016
Illustrations © Mark Janssen 2016

How Can I Help You? text © Becca Heddle 2016
Illustrations © George Bletsis 2016

How We See text © Kate Scott 2016
Illustrations © Sandra Rodriguez 2016

Dive! Dive! text © Liz Miles 2016
Illustrations © Mark Chambers 2016

The moral rights of the authors have been asserted

This Edition published in 2019

All rights reserved. No part of this publication may be reproduced, stored in a retrieval system, or transmitted, in any form or by any means, without the prior permission in writing of Oxford University Press, or as expressly permitted by law, by licence or under terms agreed with the appropriate reprographics rights organization. Enquiries concerning reproduction outside the scope of the above should be sent to the Rights Department, Oxford University Press, at the address above.

You must not circulate this work in any other form and you must impose this same condition on any acquirer

British Library Cataloguing in Publication Data
Data available

ISBN: 978-0-19-276968-8

10 9 8 7 6 5 4 3 2 1

Paper used in the production of this book is a natural, recyclable product made from wood grown in sustainable forests. The manufacturing process conforms to the environmental regulations of the country of origin.

Printed in China

Acknowledgements
Series Editor: Nikki Gamble

Big Animal Vet written by Jillian Powell

Big Animal Vet
The publisher would like to thank the following for permission to reproduce photographs: **p7**: Shutterstock; **p8**: Images of Africa Photobank/Alamy Stock Photo; **p9**: Beverly Jouber/Getty; **p10**: Tomas Dressler/Getty; **p11**: Shutterstock; **p12**: SuperStock/Alamy Stock Photo; **p14** and **p15**: Shutterstock

Exotic Plant Shop
The publisher would like to thank the following for permission to reproduce photographs: **p20**, **p27** and **p28**: iStockphoto; **p22**, **p28** and **p29**: Tamara Kulikova/Shutterstock; **p23**, **p28** and **p29**: shihina/Shutterstock; **p24** and **p28**: only_fabrizio/Bigstock; **p25** and **p28**: Peredniankina/Shutterstock; **p25**: RodneyX/Istockphoto; **p28**: Shutterstock/Pattaporn.

Things with Wings
The publisher would like to thank the following for permission to reproduce photographs: **p31**: Jelger Herder/Buiten-beeld/Getty Images; **p31**: Konstantin Kalishko/Alamy Stock Photo; **p31**: Natan Dotan/Getty Images; **p32**: Digital Camera Magazine/Getty Images; **p34**: Konstantin Kalishko/Alamy Stock Photo; **p35**: Stephen Dalton/Getty Images; **p36** and **p40**: Volodymyrkrasyuk/Dreamstime; **p37**: Kevin Elsby/Alamy Stock Photo; **p39**: Fuse/Getty Images.

All other photography supplied by Shutterstock

Off to the Beach
The publisher would like to thank the following for permission to reproduce photographs: All photos Shutterstock except: **p44**: Bill C/Bigstock; GailJohnson/Bigstock; **p45**: Pod666/Dreamstime; SusanFeldberg/Bigstock; Budda/Dreamstime; Natrow Images/Alamy; TomasSereda/iStock; **p46**: Tim Laman/Nature Picture Library; **p47**: THPStock/Bigstock; **p48**: mychadre77/ Bigstock; **p48**: Jason Edwards/ Getty Images; Nature/UIG/Getty Images; **p49**: Vspiller/iStock; **p50**: Vasiliy Vishnevskiy/Dreamstime.com; Curtsinger/Getty Images; Juniors Bildarchiv GmbH/Alamy; **p50** and **p52**: cec72/Bigstock; **p52**: zagzig/ Bigstock; Michele Westmorland/Corbis

How Can I Help You?
The publisher would like to thank the following for permission to reproduce photographs: **p55**: Shutterstock/Pattaporn; **p55** and **p56**: Shutterstock; **p57**: Hugh Chittenden; **p57**: Richard du Toit/Getty Images; **p58** and **p59**: WaterFrame/Alamy Stock Photo; **p59**: Shutterstock; **p60**: Carol Polich Photo Workshops/Getty Images; **p61**: Arief Juwono/Getty Images; Chairat/Getty Images; **p62**: Shutterstock; **p63**: Georgette Douwma/Naturepl.com; **p63b**: David Fleetham/Alamy Stock Photo; **p64**: WaterFrame/Alamy Stock Photo; Carol Polich Photo Workshops/Getty Images; Chairat/Getty Images; Shutterstock

How We See
The publisher would like to thank the following for permission to reproduce photographs: **p68**: blickwinkel/Alamy Stock Photo; **p68**, **p70**, **p72**, **p74** and **p76**: v.s.anandhakrishna/Shutterstock; **p68** and **p76**: Thomas Zsebok Images/ istock; **p69** and **p76**: Atlantide Phototravel/ Corbis; **p70**: Bagicat/Bigstock; **p70** and **p76**: Michael Thornton/ Design Pics/Getty Images; **p71**, **p76**: Macduff Everton/Getty Images; **p72**: Bildagentur Zoonar GmbH/Shutterstock; **p72** and 76: B and E. Dudzinscy/Shutterstock; **p73** and **p76**: Nickolay Lamm; **p74** and **p76**: Mariia Masich/Shutterstock; **p75** and **p76**: Saša Prudkov/Dreamstime. com

Dive! Dive!
The publisher would like to thank the following for permission to reproduce photographs: **p80**, **p81** and **p88**: Scubazoo/Alamy Stock Photo; **p82**, **p83** and **p88**: Wild Wonders of Europe/Aukan/Nature Picture Library; **p84**, **p85** and **p88**: Auscape/UIG/Getty Images; **p86**, **p87** and **p88**: Norbert Wu/Getty Images

All other images Shutterstock

Cover images Shutterstock